# STECK-VAUGHN

## Mathematics S...
### Decimals and Percents
**7700**

James T. Shea

## Contents

| | |
|---|---|
| Decimal Concepts | 2–6 |
| Rounding and Estimation | 7–8 |
| Adding Decimals | 9 |
| Subtracting Decimals | 10 |
| Applying Addition and Subtraction Skills | 11 |
| Checking Up on Adding and Subtracting Decimals | 12 |
| Multiplying Decimals | 13–14 |
| Dividing Decimals | 15–16 |
| Multiplying and Dividing by 10, 100, and 1,000 | 17 |
| Applying Multiplication and Division Skills | 18 |
| Checking Up on Multiplying and Dividing Decimals | 19 |
| Checking Up on Decimals | 20–21 |
| Percent Concepts | 22–26 |
| Understanding Percent Sentences and Equations | 27 |
| Finding the Part | 28 |
| Estimating the Part | 29–30 |
| Finding the Percent | 31 |
| Estimating the Percent | 32–33 |
| Finding the Whole | 34 |
| Estimating the Whole | 35–36 |
| Finding the Original Amount | 37 |
| Finding the Final Amount | 38 |
| Finding the Rate of Increase or Decrease | 39 |
| Simple Interest | 40 |
| Compound Interest | 41 |
| Checking Up on Percents | 42–43 |
| Progress Review | 44–45 |
| Answer Key | 46–48 |

Dr. Nancy R. Callicotte
2641 West Hidden Bluffs Drive
Tucson, AZ 85742

**STECK-VAUGHN**
ELEMENTARY · SECONDARY · ADULT · LIBRARY
A Harcourt Company

www.steck-vaughn.com

# The Meaning of Decimals

Like fractions, decimals show parts of a whole. The shaded portion of each picture can be written as a fraction or as a decimal.

 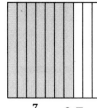 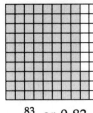 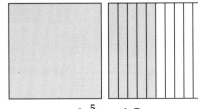

$\frac{1}{1}$ or 1    $\frac{7}{10}$ or 0.7    $\frac{83}{100}$ or 0.83    $1\frac{5}{10}$ or 1.5

one    seven tenths    eighty-three hundredths    one and five tenths

Remember,
- a decimal point separates a whole number and its decimal parts.
- a whole number has a decimal point but it is usually not written. 2 is the same as 2.0.
- when a whole number is followed by a decimal part, the decimal point is read as *and*.

**Write the decimal using numerals and words shown by the shaded part of each figure.**

1.

   __0.3__         _____          _____

   or __three tenths__   or _____     or _____

2.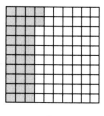

   __.32__         _____          _____

   or __32 hundredths__  or _____     or _____

3.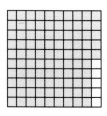

   _____        _____          _____

   or _____     or _____       or _____

Answers begin on page 46.

# Reading and Writing Decimals

To read a decimal, read as a whole number. Then name the place value of the last digit.

Read and write 0.53 as fifty-three hundredths.

To read a decimal that has a whole number part,
- read the whole number part.
- read the decimal point as *and*.
- read the decimal part as a whole number and then name the place value of the last digit.

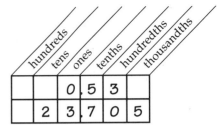

← whole number • decimal →

Read 23.705 as twenty-three and seven hundred five thousandths.

A zero placed at the end of a decimal does not change the value of that decimal. Three tenths, or 0.3, is equal to thirty hundredths, or 0.30.

**Write the following as decimals:**

1. three tenths _____ 0.3 _____
2. 25 thousandths _____
3. 15 ten-thousandths _____
4. five tenths of a yard _____
5. 15 thousandths of an inch _____
6. 20 hundredths of a ton _____
7. eighteen and nine tenths _____
8. two and one tenth _____

15 hundredths _____

five thousandths _____

three-tenths feet _____

seventy-five hundredths miles _____

4 hundredths of a mile _____

two hundred five thousandths _____

ten and sixty-five hundredths _____

seventy-one and six hundredths _____

**Write the decimals in words.**

9. 0.1 _____ one tenth _____
10. 0.06 _____
11. 0.312 _____
12. 0.5 _____
13. 0.037 _____
14. 3.49 _____
15. 17.9 _____
16. 1.40 _____

0.25 _____

0.62 _____

0.007 _____

0.16 _____

0.902 _____

35.2 _____

50.05 _____

96.621 _____

Answers begin on page 46.

# Money

The money system in the United States is based on decimals. The digits to the left of the decimal point show whole dollar amounts. The two digits to the right of the decimal point show hundredths of a dollar, or cents.

$0.29    twenty-nine cents

$32.08   thirty-two dollars and eight cents

$17.50   seventeen dollars and fifty cents

**Write each money amount as a decimal with a dollar sign and decimal point.**

1. fifty-two cents _____$0.52_____    eighteen cents _____

2. ninety-nine cents _____    seventy cents _____

3. one dollar and five cents _____    five dollars and ten cents _____

4. twelve dollars and nineteen cents _____    twenty-three dollars and fifty cents _____

5. sixty-two dollars and thirty cents _____    eighty-two dollars and nine cents _____

6. one hundred thirty-six dollars and ninety-two cents _____

**Write each money amount in words.**

7. $0.11 _____eleven cents_____    $0.80 _____

8. $0.44 _____    $0.05 _____

9. $0.09 _____    $0.71 _____

10. $5.13 _____    $7.01 _____

11. $10.10 _____    $15.22 _____

12. $39.98 _____    $52.38 _____

13. $70.03 _____    $89.47 _____

14. $123.52 _____

15. $580.50 _____

16. $427.09 _____

# Fraction and Decimal Equivalents

Sometimes you will need to either change a decimal to a fraction or a fraction to a decimal.

To write a decimal as a fraction, identify the value of the last place in the decimal. Use this place value to write the denominator. Simplify if possible.

To write a fraction that has a denominator of 10, 100, or 1,000 as a decimal, write the digits from the numerator. Write the decimal point. Notice that the number of zeros is the same as the number of decimal places in the decimal.

| Decimal | Fraction or Mixed Number |
|---------|--------------------------|
| 0.9     | = $\frac{9}{10}$ |
| 0.01    | = $\frac{1}{100}$ |
| 0.045   | = $\frac{45}{1,000} = \frac{9}{200}$ |
| 1.74    | = $\frac{174}{100} = 1\frac{74}{100} = 1\frac{37}{50}$ |

| Fraction or Mixed Number | Decimal |
|--------------------------|---------|
| $\frac{3}{10}$ | = 0.3 |
| $\frac{15}{100}$ | = 0.15 |
| $\frac{6}{1,000}$ | = 0.006 |
| $\frac{59}{10}$ or $5\frac{9}{10}$ | = 5.9 |

**Write each decimal as a fraction. Simplify if possible.**

1. 0.4  $\frac{4}{10} = \frac{2}{5}$     0.6 _____     0.08 _____     0.002 _____

2. 0.21 _____     0.083 _____     0.901 _____     0.018 _____

**Write each decimal as a mixed number. Simplify if possible.**

3. 4.5  $4\frac{5}{10} = 4\frac{1}{2}$     1.62 _____     10.1 _____     1.275 _____

4. 9.07 _____     38.24 _____     5.46 _____     13.8 _____

**Write each fraction as a decimal.**

5. $\frac{1}{10}$  .1     $\frac{2}{10}$ _____     $\frac{5}{10}$ _____     $\frac{7}{10}$ _____

6. $\frac{6}{100}$ _____     $\frac{80}{100}$ _____     $\frac{52}{1,000}$ _____     $\frac{416}{1,000}$ _____

7. $\frac{56}{10}$ _____     $\frac{31}{10}$ _____     $\frac{76}{10}$ _____     $\frac{65}{100}$ _____

8. $\frac{103}{100}$ _____     $\frac{509}{100}$ _____     $\frac{1,643}{1,000}$ _____     $\frac{2,051}{1,000}$ _____

**Answers begin on page 46.**

# Comparing and Ordering Decimals

To compare two decimals, line up the decimal points. Beginning at the left, compare the value of the digits in each place. The greater number is the number with greater digit farthest to the left.

The symbol < means *is less than*.    4.2 < 4.6
The symbol > means *is greater than*.    2.7 > 2.3
The symbol = means *is equal to*.    3.4 = 3.40

**Compare: 2.6 and 2.3**

The ones digits are the same. Compare the tenths.

6 > 3, so 2.6 > 2.3

**Compare: $0.08 and $0.25**

The ones digits are the same. Compare the tenths.

0 < 2, so $0.08 < $0.25

**Compare 0.4 and 0.47**

Write a zero. The ones and tenths digits are the same. Compare the hundredths.

0 < 7, so 0.4 < 0.47

Compare the decimals in each pair. Write <, >, or =.

1. 0.3 _=_ 0.30        0.5 ____ 0.500        0.035 ____ 0.35        0.25 ____ 0.3

2. 3.5 ____ 3.50        4.50 ____ 4.500        0.125 ____ 13        0.15 ____ 0.115

3. 0.625 ____ 0.6250        0.035 ____ 0.0350        0.26 ____ 0.3        0.6 ____ 0.65

Write these numbers in order, beginning with the smallest:

4. $25  $2.50  $0.25  $1.25  $1.02  $1.20  $1.10  $1.01 _____

If the numbers in each pair below are the same, write *S* on the line. If they are not the same, write *D* on the line:

5. One hundred twenty-five — 0.125 _D_        Three fourths — 0.34 _____

6. One and six tenths — 1.6 _____        Four twenty-fifths — 4.25 _____

7. Thirty-five and one half — 35.2 _____        Eight thousandths — 0.008 _____

8. Ten and one fifth — 10.5 _____        Five and one tenth — 5.10 _____

9. $\frac{1}{10}$ ____ 0.01        $\frac{5}{10}$ ____ 0.5        $\frac{90}{10}$ ____ 0.9        $\frac{2}{10}$ ____ 2.0

10. $\frac{7}{100}$ ____ 0.7        $\frac{24}{100}$ ____ 0.24        $\frac{1}{100}$ ____ 0.01        $\frac{130}{100}$ ____ 0.13

11. $\frac{82}{100}$ ____ 0.082        $\frac{33}{100}$ ____ 0.003        $\frac{60}{100}$ ____ 0.6        $\frac{4}{100}$ ____ 0.4

12. $\frac{9}{1,000}$ ____ 0.09        $\frac{42}{1,000}$ ____ 0.42        $\frac{500}{1,000}$ ____ 0.5        $\frac{110}{1,000}$ ____ 1.1

13. The last digit on all speedometers expresses tenths. The speedometer on Clark's car has five 9's. Express this mileage in decimal form. _____

14. When his car goes another one tenth of a mile, what will the speedometer read? _____

# Rounding Decimals

To round a decimal to the nearest whole number, look at the digit in the tenths place. If the digit is 5 or greater, add 1 to the number in the ones place. If the digit is less than 5, drop all the digits to the right of the decimal point. When rounding money amounts you may leave zeros after the decimal point as place holders.

    3.2 rounds down to 3        $37.81 rounds up to $38.00
    0.529 rounds up to 1         $147.06 rounds down to $147.00

**Round to the nearest whole number or nearest dollar.**

1. 5.7 __6__   2.1 _____   0.9 _____   314.6 _____   18.5 _____

2. $27.62 __$28.00__   $519.93 _____   $182.67 _____   $92.08 _____   $5.52 _____

3. 244.036 _____   1.099 _____   0.103 _____   118.927 _____   129.335 _____

To round a decimal to the nearest tenth, look at the digit in the hundredths place. If the digit is 5 or greater, add 1 to the number in the tenths place. If the digit is less than 5, drop all the digits to the right of the tenths place. For money amounts, add a zero as a place holder in the hundredths place.

    0.59 rounds up to 0.6        $0.79 rounds up to $0.80
    1.327 rounds down to 1.3    $9.51 rounds down to $9.50

**Round to the nearest tenth or nearest dime.**

4. $0.68 __$0.70__   $0.51 _____   $2.83 _____   $14.95 _____   $32.09 _____

5. 29.26 __29.3__   33.33 _____   92.07 _____   5.39 _____   15.55 _____

6. 12.245 _____   0.625 _____   $166.666 _____   $1.119 _____   423.833 _____

To round a decimal to the nearest hundredth, look at the digit in the thousandths place. If the digit is 5 or greater, add 1 to the number in the hundredths place. If the digit is less than 5, drop all the digits to the right of the hundredths place.

    0.767 rounds up to 0.77      $0.455 rounds up to $0.46
    3.9243 rounds down to 3.92   $1.021 rounds down to $1.02

**Round to the nearest hundredth or nearest cent.**

7. 1.527 __1.53__   3.035 _____   0.375 _____   8.666 _____   0.333 _____

8. $10.075 __$10.08__   $18.276 _____   $5.899 _____   $27.987 _____   $1.329 _____

9. 435.3275 _____   $0.5543 _____   $11.0981 _____   183.2678 _____   136.3725 _____

Answers begin on page 46.

# Estimating with Decimals

Estimation can be used to predict the exact answer or to check the reasonableness of an answer. To estimate with decimals, first round to the nearest whole number, tenth, or hundredth. Then solve the problem.

**Estimate: $3.74 + $5.37**

| Round each decimal to the nearest whole dollar. Then add. | Round each decimal to the nearest tenth (dime) to get a more accurate estimate. Then add. | Find the exact answer. |
|---|---|---|
| $3.74 → $4.00<br>+ 5.37 → + 5.00<br>$9.00 | $3.74 → $3.70<br>+ 5.37 → + 5.40<br>$9.10 | $3.74<br>+ 5.37<br>$9.11 |

**Estimate by rounding to the nearest whole number or nearest dollar. Then find the exact answer.**

1. 
   7.8 → 8     7.8     $43.16           31.225
   +12.1 → +12  +12.1   − 29.88          ×14.677       30)119.7
           20    19.9

**Estimate by rounding to the nearest tenth or nearest dime. Then find the exact answer.**

2. 
   $9.37 → $9.40     87.92      $127.40
   × 0.51 → ×  .50   + 13.674   −  97.86         12)198.33
           $4.70

3. 
   0.255        93.470     $1.099
   − 0.16       + 8.699    ×    10            22)24.244

**Solve.**

4. Gareth bought 15.8 gallons of gas. The pump price was $1.099 per gallon. Which expression is the best estimate of what he paid for the gas?

   a) 15 × $1.00
   b) 15 ÷ $1.10
   c) 16 × $1.10

5. A 7-pound bag of cat food costs $3.85. There are 3.18 kilograms in 7 pounds. Which is the best estimate of the cost per kilogram?

   a) $3.00 ÷ 3.2
   b) $4.00 ÷ 3
   c) $4.00 × 3

# Adding Decimals

To add decimals, line up the decimal points. Write zeros as placeholders and rename if necessary. Then add as with whole numbers. Be sure to write a decimal point in your answer.

Find: 4.6 + 7.32

| Write a zero as a placeholder. | Add the hundredths. | Add the tenths. Write a decimal point in the answer. | Add the ones. |
|---|---|---|---|
| T\|O\|Ts\|Hs<br>   4 .6  0<br>+ 7 .3  2 | T\|O\|Ts\|Hs<br>   4 .6  0<br>+ 7 .3  2<br>             2 | T\|O\|Ts\|Hs<br>   4 .6  0<br>+ 7 .3  2<br>      .9  2 | T\|O\|Ts\|Hs<br>   4 .6  0<br>+ 7 .3  2<br> 1 1 .9  2 |

**Add. Write zeros as placeholders if needed.**

1.  
$\overset{1}{\$5}.95$  
$+\ 3.62$  
$\overline{\ \$9.57\ }$

   $\$2.60$  
  $+\ 6.75$

   $\$0.08$  
  $+\ 0.26$

   $14.07$  
  $+\ \ 8.46$

   $17.05$  
  $+\ \ 3.35$

2.  
$28.009$  
$+\ \ 6.003$

   $6.126$  
  $+\ 0.04$

   $77.016$  
  $+\ \ 0.647$

   $84.70$  
  $+\ \ 3.08$

   $8.23$  
  $+0.009$

3.  
$0.15$  
$0.6$  
$+3$

   $3$  
   $0.07$  
  $+7.325$

   $0.009$  
   $0.6$  
  $+2.10$

   $\$0.14$  
   $0.06$  
  $+\ \ 8.50$

   $0.005$  
   $0.034$  
  $+7.16$

**Line up the decimal points and add. Write zeros as placeholders if needed.**

4. $12.20 + $2.10 =       $7 + $3.22 =       $0.50 + $0.92 =

5. $5.95 + $14.56 + $22.06 =       $15.29 + $37.70 + $96.08 =

**Estimate by rounding to the nearest hundredth or cent. Then find the exact answer. Write zeros as placeholders if needed.**

6.  
$\ \ \ \ 7.368\ \rightarrow\ \ \ \ \ \ \ 7.37$  
$+159.93\ \rightarrow\ +159.93$  
$\ \ \ \ \ \ \ \ \ \ \ \ \ \ \ \ \ \ \ \ \ \overline{167.30}$

$\ \ \ \ \ \ \ \ \ \ \ \ \ \ \ \ \ \ \ \ \ 7.368$  
$\ \ \ \ \ \ \ \ \ \ \ \ \ \ \ +159.930$  
$\ \ \ \ \ \ \ \ \ \ \ \ \ \ \ \overline{167.298}$

   $\$198.892$  
  $+\ \ 11.935$

   $\$1.991$  
  $+\ 6.39$

   $\$0.445$  
  $+\ 0.135$

Answers begin on page 46.

# Subtracting Decimals

To subtract decimals, line up the decimal points. Add zeros as placeholders so that both decimals have the same number of digits after the decimal point. Then subtract as with whole numbers. Regroup if necessary. Be sure to write a decimal point in your answer.

Find: 34.3 − 17.94

| Write a zero as a placeholder. Regroup to subtract the hundredths. | Regroup to subtract the tenths. Write a decimal point in the answer. | Regroup to subtract the ones. | Subtract the tens. |
|---|---|---|---|
| T\|O\|Ts\|Hs<br>  \|  \| 2 \|10<br>3\|4\|.3\|0̸<br>−1\|7\|.9\|4<br>  \|  \|   \| 6 | T\|O\|Ts\|Hs<br>  \|  \|12\|<br>  \|3\|2̸\|10<br>3\|4̸\|.3̸\|0̸<br>−1\|7\|.9\|4<br>  \|  \|.3\|6 | T\|O\|Ts\|Hs<br>  \|13\|12\|<br>2\|3̸\|2̸\|10<br>3̸\|4̸\|.3̸\|0̸<br>−1\|7\|.9\|4<br>  \|6\|.3\|6 | T\|O\|Ts\|Hs<br>  \|13\|12\|<br>2\|3̸\|2̸\|10<br>3̸\|4̸\|.3̸\|0̸<br>−1\|7\|.9\|4<br>1\|6\|.3\|6 |

**Subtract. Write zeros as placeholders if needed.**

1.  $\$8.32$         $\$8.99$         $\$7.27$         $9.068$         $10.399$
    $-\ 3.20$       $-\ 4.25$        $-\ 5.12$        $-7.054$        $-10.239$
    $\$5.12$

2.  $\$34.95$        $\$0.62$         $\$92.00$        $\$94.78$       $\$11,532.30$
    $-\ 27.99$      $-\ 0.57$        $-\ 67.50$       $-\ 15.00$      $-\ \ \ 2,500.00$

3.  $\overset{\ \ \ \ \ \ 11}{\underset{\ }{5\ \cancel{1}\ 9\ 10}}$
    $\cancel{6.200}$   $\$10.00$      $5$              $\$1.35$        $\$100.00$
    $-4.575$        $-\ \ 6.13$      $-0.674$         $-\ 0.26$        $-\ \ 84.7$
    $1.625$

**Line up the decimal points and subtract. Write zeros as placeholders if needed.**

4.  $8.09 - 4.256 =$          $\$5 - \$3.49 =$          $\$85.49 - \$12 =$

5.  $\$10.00 - \$4.50 =$       $1.52 - 0.48 =$           $\$0.90 - \$0.37 =$

**Estimate by rounding to the nearest tenth or nearest dime. Then find the exact answer.**

6.  $\$92.15 \rightarrow \$92.20$     $4.78$        $\$20.00$       $5.02$
    $-\ \ \ 8.47 \rightarrow -\ \ 8.50$   $-0.59$       $-\ \ 6.98$      $-1.863$
    $\ \ \ \ \ \ \ \ \ \ \ \ \ \ \ \$83.70$

    $\ \ \ \ \ \ \ \ \ \ \ \ \ \ \ \$92.15$
    $\ \ \ \ \ \ \ \ \ \ \ \ -\ \ \ 8.47$
    $\ \ \ \ \ \ \ \ \ \ \ \ \ \ \ \$83.68$

Answers begin on page 46.

# Applying Addition and Subtraction Skills

Estimate. Then find the exact answer.

1. Trudy's lunch at Casa Ramirez was $4.95. If she paid $0.29 in tax and left $.95 for a tip, what was the total cost of her lunch?

   Estimate _____ Answer _____

2. On the average, it takes a worker on a certain assembly line 0.86 minutes to get a part from a bin and 4.91 minutes to attach the part to the engine being made. In all, how long does it take to get the part and attach it?

   Estimate _____ Answer _____

3. Jack paid a total of $5,695.25 for his car. If $470.25 of the total was for interest on his car loan, what was the price of the car?

   Estimate _____ Answer _____

4. When Jack bought his car, the speedometer read 98,476.3. After one month, it read 100,026.7. How many miles did Jack drive during the month?

   Estimate _____ Answer _____

5. At Townsend's Tools, workers record each quarter hour they work as 0.25 hours. If Lou recorded 5.25 hours on Monday, 7.50 hours on Tuesday, and 6.75 hours on Wednesday, how many hours did he work in all?

   Estimate _____ Answer _____

6. The diagonal of Bart's new patio should measure about 13.453 feet. But after it was built Bart found that the diagonal measured $13\frac{1}{2}$ feet. How much longer is the diagonal than it should be? (Hint: Change $13\frac{1}{2}$ feet to a decimal.)

   Estimate _____ Answer _____

7. In 1992, Carl Lewis held the men's world record, 9.86 seconds, for the 100 meter race. Florence Griffith-Joyner held the women's world record, 10.49 seconds, for the 100 meter race. How many seconds faster was Carl Lewis than Florence Griffith-Joyner?

   Estimate _____ Answer _____

8. After a paperback book is printed and bound, the edges are trimmed so that they are even. If a book is printed on 7.5 inch-wide paper, and $\frac{7}{8}$, or 0.875, of an inch is trimmed, what is the final width of the book?

   Estimate _____ Answer _____

9. Bluebonnet Electric Company states that their customers pay an average of $68.07 per 1,000 kilowatt hours. The state average is $76.61. How much less is Bluebonnet's average than the state average?

   Estimate _____ Answer _____

10. The two top batters in the American League in 1996 had batting averages that differed by 0.014. If the number two player batted 0.329, what was the batting average of the number one player?

    Estimate _____ Answer _____

Answers begin on page 46.

# Checking Up on Adding and Subtracting Decimals

Write the following in words.

1. 3.29 _____   17.006 _____

2. $14.50 _____   $0.67 _____

Write each decimal as a fraction or mixed number. Write each fraction as a decimal.

3. 0.9 _____   4.01 _____   $\frac{8}{10}$ _____   $\frac{130}{100}$ _____

Compare. Write <, >, or =.

4. $\frac{4}{10}$ _____ 0.04   0.125 _____ 0.1250   0.637 _____ 0.649   $\frac{82}{1,000}$ _____ 0.009

Round as indicated.

5. 24.09 (nearest whole number) _____   $7.99 (nearest dollar) _____

6. $0.375 (nearest cent) _____   18.547 (nearest tenth) _____

7. 0.5213 (nearest hundredth) _____   $93.89 (nearest dime) _____

Estimate by rounding as indicated. Then find the exact answer.

8. (nearest tenth)   3.09 + 0.87 =

   (nearest dollar)   $40 − $18.79 =

   (nearest dime)   $1.29 + $12 =

9. (nearest hundredth)   0.682 + 1.204 =

   (nearest cent)   $1.079 + $10 =

   (nearest whole number)   162.03 − 17.9 =

Add or subtract. Use zeros as place holders if needed.

10.  $0.68         3.7          $19.42         0.276          $527.83
     0.32          1.45          23.88         3.109            49.62
   + 1.79        + 2.06        +  8.79       + 5.081         + 113.40


11.   15.03        $42.80       $360.00         1.321         $4,825.50
    − 1.067      − 17.36       −  92.75       −0.058         −   967.29


12. $20 − $11.93 =        $43.20 + $1.93 =        $50 − $43.20 =


13. $6.98 + $14.38 + $9.90 =        $5 − $1.67 =        $7.05 + $13.95 =

# Multiplying Decimals and Whole Numbers

To multiply decimals and whole numbers, multiply as if you were multiplying whole numbers. Count the numbers of decimal places to the right of the decimal point in the numbers you have multiplied. The product will have the same number of decimal places. Place the decimal point in the product.

Remember, sometimes you might need to write a zero in the product in order to place the decimal point correctly.

**Find: 13 × 6.4**

Multiply. Write the decimal point in the product.

```
     6.4   ← 1 decimal place
   ×13     ← +0 decimal places
   ─────
    192
    64
   ─────
    83.2   1 decimal place
```

**Find: 0.018 × 5**

Multiply. Write the decimal point in the product.

```
        5   ← 0 decimal places
   ×0.018   ← +3 decimal places
   ──────
    0.090     3 decimal places
         └── Write a zero.
```

**Multiply. Write zeros as needed. Round answers in money problems to the nearest cent.**

1. 
   $$0.862 \times 2 = 1.724 \qquad 0.084 \times 3 \qquad \$1.63 \times 6 \qquad 3.6 \times 4 \qquad \$2.34 \times 5$$

2. 
   $$\$13.60 \times 2 \qquad \$28.52 \times 4 \qquad \$11.30 \times 3 \qquad \$33.34 \times 3 \qquad \$26.00 \times 5$$

3. 
   $$8.2 \times 12 \qquad 362 \times 0.35 \qquad \$70 \times 5.0 \qquad \$1.35 \times 0.15 \qquad 0.90 \times 55$$

4. $\$3.07 \times 25 =$ \qquad $234 \times 0.05 =$ \qquad $0.045 \times 82 =$

**Estimate by rounding to the nearest dollar or whole number. Then find the exact answer.**

5. 
   $$6.01 \times 24 \qquad 0.891 \times 332 \qquad \$5.25 \times 60 \qquad \$175 \times 1.8$$

Answers begin on page 46.

# Multiplying Decimals by Decimals

To multiply decimals by decimals, multiply as if you were multiplying whole numbers. Place the decimal in the answer by counting the number of decimal places to the right of the decimal points in both numbers you are multiplying. The answer will have the same number of decimal places. Write zeros as needed.

**Find: 0.48 × 13.7**

| Multiply. Write the decimal point in the product. |
|---|
| 13.7 ← 1 place<br>× 0.48 ← + 2 places<br>―――<br>1096<br>548<br>―――<br>6.576      3 places |

**Find: 0.008 × 0.137**

| Multiply. Write the decimal point in the product. |
|---|
| 0.137 ← 3 places<br>× 0.008 ← + 3 places<br>―――<br>0.001096     6 places<br>       Write 2 zeros. |

**Multiply. Write zeros as needed. Round answers in money problems to the nearest cent.**

1.  0.3        0.03       $5.60       $9.84        2.5        $6.30
   × 0.6      × 0.6      × 0.2       × 0.5      × 0.03      × 0.04
   ――――
   0.18

2. $0.08       0.09       0.006       0.008       $1.37       $1.37
   × 0.4      ×0.09      × 0.4       × 0.06      × 0.8      ×0.008

3. 0.015       0.075      $0.325      $15.50       10.9       $7.05
   × 0.14     × 0.22     × 0.19      × 0.12      ×0.34      × 0.54

4. $25.50 × 1.5 =          $16.37 × 0.085 =          $79.95 × 0.15 =

**Estimate by rounding to the nearest tenth or nearest dime. Then find the exact answer.**

5. $14.70          $21.35          520.03          0.932
   × 0.09         × 0.39          × 0.75         × 0.516

Answers begin on page 46.

# Dividing Decimals by Whole Numbers and Decimals

To divide decimals, put the decimal point in the answer directly above the decimal point in the problem. Then divide as with whole numbers.

If the divisor is a decimal, move the decimal point to the right to make a whole number. Move the decimal in the dividend the same number of places. Then divide.

**Find: 9.68 ÷ 16**

| Write a decimal point in the answer. | Divide. Insert a zero in the answer. |
|---|---|
| .<br>16)9.68 | 0.605<br>16)9.680<br>   96↓↓<br>     80<br>     80<br>      0 |

**Find: $48 ÷ 0.24**

| Move the decimal points two places. Insert zeros if needed. | Divide. |
|---|---|
| 0.24)$48.00 | $200<br>24)$4800<br>   48↓↓<br>     00<br>     00<br>      0 |

**Divide. Round money amounts to the nearest cent.**

1.     8.2
    8)65.6
      64
      16
      16
       0
    
    5)$3.45     3)$8.28     7)0.784

2.    0.04
    61)2.44
      244
        0
    
    39)$58.50     46)9.338     14)$43.96

3.    41.6
    0.7)29.12
      28
      11
       7
       42
    
    0.4)$16.48     0.71)1.278     0.22)0.154

4. $26.50 ÷ 0.5 =     1.92 ÷ 0.16 =     $1.68 ÷ 0.14 =

**Estimate by rounding to the nearest tenth or nearest dime. Then divide.**

5. 2)$17.39     3.9)3.89     0.45)$90.93     0.15)3014

# Dividing Smaller Numbers by Larger Numbers

Sometimes when you divide, the divisor will be larger than the dividend. To divide, add a decimal point and zeros as needed to the dividend. Continue to divide until the remainder is zero. In some cases, you may never have a remainder of zero. Then you have to round the answer to a certain place. When dividing money, round the answer to the nearest cent. You may need to add zeros to the answer also.

To change fractions to decimals you can use this method. Divide the numerator by the denominator.

Find: $19 ÷ 300                           Change $\frac{8}{125}$ to a decimal.

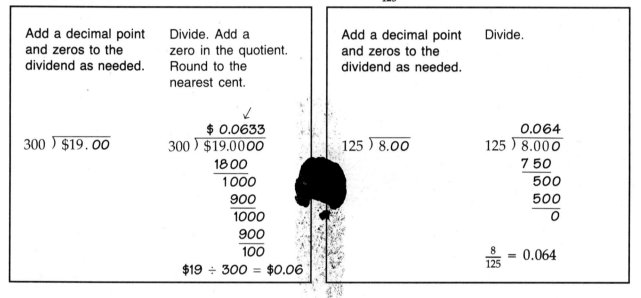

Divide. Write zeros as needed. Round repeating decimals to the nearest hundredth or cent.

1.  
   $40\overline{)20.0}$ quotient 0.5, 200, 0
   
   $100\overline{)\$30}$    $5\overline{)\$2}$    $12\overline{).24}$    $20\overline{)\$15}$

2. $50\overline{)1}$    $500\overline{)25}$    $3\overline{)\$2}$    $4\overline{)\$1}$    $20\overline{)\$19}$

Use division to change each fraction to a decimal. Round repeating decimals to the nearest tenth.

3. $\frac{1}{4}$    $\frac{1}{8}$    $\frac{1}{3}$    $\frac{3}{5}$    $\frac{3}{16}$

4. $\frac{6}{12}$    $\frac{5}{8}$    $\frac{1}{6}$    $\frac{4}{9}$    $\frac{3}{4}$

# Multiplying and Dividing by 10, 100, and 1,000

To multiply by 10, 100, or 1,000 move the decimal point in the answer to the right as many places as there are zeros in the multiplier.

You may need to write zeros in the answer in order to move the decimal point the correct number of places.

| 10 × 0.89 = 8.9 | 100 × 0.73 = 73 | 1,000 × 0.52 = 520 |
| 10 × 8.9 = 89 | 100 × 7.3 = 730 | 1,000 × 5.2 = 5,200 |

To divide a decimal by 10, 100, or 1,000 move the decimal point in the dividend to the left as many places as there are zeros in the divisor.

You may need to write zeros in the answer in order to correctly insert the decimal point the correct number of places.

| 0.89 ÷ 10 = 0.089 | 0.73 ÷ 100 = 0.0073 | 0.52 ÷ 1,000 = 0.00052 |
| 8.9 ÷ 10 = 0.89 | 7.3 ÷ 100 = 0.073 | 5.2 ÷ 1,000 = 0.0052 |

**Multiply.**

1. 7.5 × 10 = ___75___      $46 × 10 = _____      0.07 × 10 = _____

2. 100 × 0.7 = _____      $100 × 4.6 = _____      0.075 × 100 = _____

3. 0.5 × 10 = _____      0.8 × 1,000 = _____      $1.25 × 100 = _____

4. 12.5 × 100 = _____      0.125 × 1,000 = _____      $14.92 × 100 = _____

5. 6.2 × 1,000 = _____      $642.15 × 10 = _____      $642.15 × 100 = _____

**Divide.**

6. $3.15 ÷ 1,000 = ___$0.00315___      0.048 ÷ 100 = _____      0.048 ÷ 1,000 = _____

7. 0.375 ÷ 10 = _____      $3.75 ÷ 10 = _____      37.5 ÷ 10 = _____

8. $375 ÷ 10 = _____      0.375 ÷ 1,000 = _____      0.007 ÷ 1,000 = _____

9. $719.35 ÷ 100 = _____      16.147 ÷ 1,000 = _____      $14.92 ÷ 1,000 = _____

10. $267.18 ÷ 100 = _____      2.6718 ÷ 1,000 = _____      2.6718 ÷ 100 = _____

Answers begin on page 46.

# Applying Multiplication and Division Skills

Solve.

1. Written as a decimal, the self employment social security tax is 0.153. If a babysitter makes $10,000.00 in a year, how much social security tax will the babysitter pay?

   Answer _____

2. One of the new commuter trains travels 436.5 miles in 3 hours. What is its average hourly speed?

   Answer _____

3. Mr. Stone paid $63.56 for a bus ticket from Chicago to New York. If the cost per mile is $0.07, what distance did he travel?

   Answer _____

4. A gallon of water weighs 8.355 pounds. If a gallon of milk weighs 1.03 times as much, how much does the milk weigh?

   Answer _____

5. Maria Torres lives 6.4 miles from work. She drives her car to and from work each day. How far does she drive in a 5-day week?

   Answer _____

6. One mile equals 1.6093 kilometers. How many kilometers does Maria drive in 5 days?

   Answer _____

7. The sales tax in Texas is $8\frac{1}{2}$ percent, or 0.085 as a decimal. How much is the sales tax on a purchase of $100.00?

   Answer _____

8. Red's Real Estate Company has a 76.8-acre plot of land. They wish to divide it into 24 lots. If divided equally, what will be the size of each lot?

   Answer _____

9. Maxine drove her taxicab 249.7 miles on 11 gallons of gas. How many miles did she average to the gallon?

   Answer _____

10. Raymond Lindell raised 355.1 bushels of corn on 13.4 acres. What was the average yield in bushels per acre?

    Answer _____

# Checking Up on Multiplying and Dividing Decimals

**Multiply. Write zeros as needed. Round answers in money problems to the nearest cent.**

1. $0.99 × 6        2.9 × 0.39        $5.49 × 0.75        1.327 × 0.516        $29.07 × 8

2. $0.29 × 15       1.8 × 1.4         $11.19 × 24         $209 × 0.15          3.8 × 0.29

**Divide. Write zeros as needed. Round answers to the nearest cent or hundredth.**

3. 6)$3.48          9)0.981           0.5)1.56            3)$7                 0.4)18

4. 12)$32.50        50)$25            1.7)1615            0.43)29.7

**Estimate. Then find the exact answer. Round answers to the nearest cent or hundredth.**

5. $0.82 × 0.09     $59.23 × 0.81     8)$31.76            1.9)2.3

**Multiply or divide.**

6. $39.50 ÷ 10 =        1.07 × 100 =        0.006 × 1,000 =

7. 1.126 × 10 =         4.9 ÷ 100 =         1.23 × 1,000 =

Answers begin on page 46.

# Checking Up on Decimals

**Write the decimals in words.**

1. 0.01 _____    3.24 _____

2. $25.30 _____    $0.19 _____

**Write each decimal as a fraction or mixed number. Write each fraction as a decimal.**

3. 0.23 _____   $\frac{4}{10}$ _____   6.09 _____   $\frac{325}{100}$ _____

**Compare. Write <, >, or =.**

4. 0.36 _____ 0.306    $\frac{1}{10}$ _____ 0.001    $\frac{147}{1,000}$ _____ 1.47    8.09 _____ 8.090

**Round as indicated.**

5. 32.15 (nearest whole number) _____    $42.99 (nearest dollar) _____

6. $0.426 (nearest cent) _____    7.352 (nearest tenth) _____

7. 0.4780 (nearest hundredth) _____    $8.45 (nearest dime) _____

**Estimate by rounding. Then find the exact answer.**

8. (nearest tenth)   4.86 − 2.75 =    9. (nearest hundredth)   0.523 × 8 =

   (nearest dollar)   $18.92 + $37.06 =       (nearest cent)   $2.045 − $1.09 =

   (nearest dime)   $3.47 × 10 =            (nearest whole number)   5.7)257.982

**Add, subtract, multiply, or divide. Round answers to money problems to the nearest cent.**

10. $7.94      $5.37      $1.25      $7.76       0.8)9.6
    − 4.56     + 1.92     × 0.04      6.67
                                      + 4.39

11. 14.027     3.08       0.506      6)$3        5.45
    + 3.6      × 12       −0.1892                − 2.9

12. 12)1.44    $306.09    $0.48      0.05        6)0.36
               − 46.45    − 0.02     × 0.09

# Checking Up on Decimals

**13.**
$2.08
  0.45
+ 3.7

    35
×  2.4

0.12)144

  1.923
  2.749
+ 1.637

  7.4763
− 6.4767

**14.**
  2.075
× 0.08

  0.018
+ 0.209

  0.992
− 0.978

  $30.29
×   0.15

3.2)480

**15.** 3.6 × 10 =         $29.30 × 100 =         1.6 ÷ 10 =

**16.** $0.09 × 1,000 =    $5.95 ÷ 100 =          246 ÷ 1,000 =

**17.** 0.001 × 10 =       $461 ÷ 10 =            $0.32 × 100 =

**Estimate. Then find the exact answer.**

**18.** Mr. Lee and his family drove to New Orleans and back, a distance of 8,872.5 miles, using 455 gallons of gas on the trip. What was their mileage per gallon?

Estimate _____  Answer _____

**19.** Diane worked 38.6 hours last week. At $4.55 per hour, how much did she earn?

Estimate _____  Answer _____

**20.** Maggie bought a stereo for $399.95. She paid $23.90 in sales tax. How much did she pay in all?

Estimate _____  Answer _____

**21.** Angus shops at a grocery store where he gets a senior-citizen discount of 10%. Before the discount his bill was $35.62. After the discount it was $32.06. How much did he save using the discount?

Estimate _____  Answer _____

**22.** A copper wire 48 feet long was divided into 15 pieces. How long was each piece?

Estimate _____  Answer _____

**23.** Ted's farm produces 128.6 pounds of butter a week. How much will it produce, at the same rate, in a year (52 weeks)?

Estimate _____  Answer _____

Answers begin on page 46.

# The Meaning of Percents

Percent means hundredths, per hundred, or part of one hundred. When using percents, the whole is divided into 100 equal parts. The symbol % is read as *percent*.

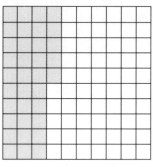

The large square is divided into one hundred smaller squares. Thirty-five of the one hundred smaller squares, or $\frac{35}{100}$, or 35% are shaded. Sixty-five of the one hundred smaller squares, or $\frac{65}{100}$, or 65% are unshaded. The total of the shaded and unshaded parts is 100%.

$35\% = \frac{35}{100} = 0.35$

Write the fraction and the percent of smaller squares shaded in each large square.

1.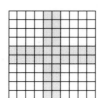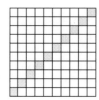

$\frac{5}{100}$ or 5%      ___ or ___      ___ or ___      ___ or ___

Write the percent shaded, the percent unshaded, and the total percent.

2.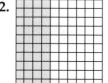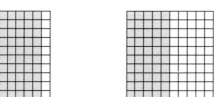

40 % shaded,
60 % unshaded,
100% total

Shade each square as indicated. Write the percent as a fraction.

3. 5%          95%          27%          75%

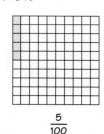 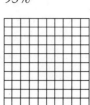  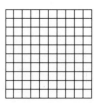

$\frac{5}{100}$

Solve.

4. A dollar is 100 cents. Write a fraction and percent of one dollar for each of the following.

   $0.05          $0.25          $0.50          $0.10          $1.00
   
   $\frac{5}{100} = 5\%$

Answers begin on page 46.

# The Meaning of Percents

Percents can be larger than 100% or smaller than 1%.

150% means $\frac{150}{100}$

200% means $\frac{200}{100}$

375% means $\frac{375}{100}$

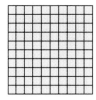 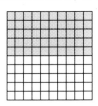

100% + 50% = 150%

$\frac{1}{2}$% or 0.5% means $\frac{0.5}{100}$ or $\frac{5}{1,000}$

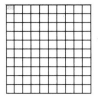

$\frac{1}{2}$ of 1% = 0.5%

Notice that 0.5% is not the same as 0.5 because 0.5 means 50%.

**Write the percents in order from smallest to largest.**

1. 1%, 10%, 0.1%, 100% _____ 0.1%, 1%, 10%, 100% _____

2. 0.5%, 5%, 50%, 500% _____

3. 2.5%, 25%, 250%, 0.25% _____

4. 75%, 7.5%, 0.75%, 750% _____

**Match each fraction with an equivalent percent.**

5. $\frac{5}{100}$ = __5%__            2.5%

6. $\frac{100}{100}$ = _____          50%

7. $\frac{10}{100}$ = _____           150%

8. $\frac{150}{100}$ = _____          10%

9. $\frac{50}{100}$ = _____           5%

10. $\frac{25}{1,000}$ = _____        100%

Answers begin on page 46.

# Changing Decimals and Fractions to Percents

To change a decimal to a percent move the decimal point 2 places to the right (or multiply by 100) and write a percent symbol (%). Add zeros as needed.

| 0.825 = 82.5%  0.03 = 3%  0.4 = 40%  1.5 = 150%  0.001 = 0.1% |
|---|

**Write each decimal as a percent.**

1. 0.30 = __30%__    0.008 = _____    2.45 = _____    3.75 = _____

2. 0.91 = _____    1.5 = _____    0.015 = _____    0.0004 = _____

3. 0.73 = _____    0.672 = _____    0.02 = _____    0.67 = _____

4. 3.25 = _____    0.09 = _____    0.7 = _____    23.93 = _____

To change a fraction to a percent, first change the fraction to a decimal by dividing the numerator by the denominator. Then rewrite the decimal quotient as a percent.

**Write $\frac{3}{4}$ as a percent.**

Divide.

$$4\overline{)3.00} = 0.75 = 75\%$$

**Write $\frac{3}{20}$ as a percent.**

Divide.

$$20\overline{)3.00} = 0.15 = 15\%$$

**Write $\frac{1}{3}$ as a percent.**

Divide. Notice that the quotient is a repeating decimal. Write the remainder as a fraction.

$$3\overline{)1.0000} = 0.3333 = 33\frac{1}{3}\%$$

**Write each fraction as a percent.**

5. $\frac{3}{8}$ = __37.5%__    $\frac{10}{5}$ = _____    $\frac{7}{10}$ = _____

6. $\frac{21}{100}$ = _____    $\frac{50}{16}$ = _____    $\frac{1}{4}$ = _____

7. $\frac{3}{5}$ = _____    $\frac{17}{20}$ = _____    $\frac{2}{1,000}$ = _____

8. $\frac{2}{3}$ = _____    $\frac{1}{9}$ = _____    $\frac{5}{12}$ = _____

# Changing Percents to Decimals and Fractions

To change a percent to a decimal, move the decimal point two places to the left (or divide by 100) and drop the percent sign (%). Write zeros as placeholders as needed.

$$35\% = 0.35 \qquad 6\% = 0.06 \qquad 480\% = 4.8 \qquad 0.2\% = 0.002$$

To change a percent to a fraction, place the percent over 100 and drop the % sign. Simplify.

$$40\% = \frac{40}{100} = \frac{4}{10} = \frac{2}{5} \qquad 150\% = \frac{150}{100} = 1\frac{1}{2} \qquad 6\% = \frac{6}{100} = \frac{3}{50}$$

$$\frac{1}{2}\% = 0.5\% = \frac{0.5}{100} = \frac{5}{1,000} = \frac{1}{200} \qquad 7\frac{1}{2}\% = \frac{7.5}{100} = \frac{75}{1,000} = \frac{3}{40}$$

**Change each percent to a decimal and then to a fraction. Simplify.**

1. $82\% =$ _____ $0.82 = \frac{82}{100} = \frac{41}{50}$ _____    $7\% =$ _____

2. $1\% =$ _____    $142\% =$ _____

3. $9\frac{1}{2}\% =$ _____    $55\% =$ _____

4. $109\% =$ _____    $73\% =$ _____

5. $98\% =$ _____    $\frac{2}{5}\% =$ _____

6. $4\% =$ _____    $44\% =$ _____

7. $175\% =$ _____    $83\% =$ _____

8. $26\% =$ _____    $0.5\% =$ _____

9. $0.3\% =$ _____    $127\% =$ _____

10. $12\frac{1}{2}\% =$ _____    $95\% =$ _____

11. $0.05\% =$ _____    $2,000\% =$ _____

12. $6\frac{1}{2}\% =$ _____    $500\% =$ _____

Answers begin on page 46.

# Fraction, Decimal, and Percent Equivalents

**Fill in the blanks below as illustrated in the first example.**

1. $\frac{1}{10}$ = 0.1 = 10%    7. $\frac{300}{100}$ = _____ = _____

2. _____ = 0.65 = _____    8. _____ = 1.9 = _____

3. _____ = _____ = 23%    9. _____ = _____ = 250%

4. $\frac{17}{20}$ = _____ = _____    10. $\frac{1}{3}$ = _____ = _____

5. _____ = 0.5 = _____    11. _____ = $0.66\frac{2}{3}$ = _____

6. _____ = _____ = 75%    12. _____ = _____ = $\frac{1}{2}$%

**Solve.**

13. Mei Ling earns $1,600 a month. She spends $350 per month on rent. What percent of her income does she spend on rent?

    Answer _____

14. The sales tax in one city is $7\frac{1}{2}$%. Write $7\frac{1}{2}$% as a decimal.

    Answer _____

15. A suit is on sale for 10% off the regular price. Write the decimal the salesperson would use to figure the amount of the discount.

    Answer _____

16. At Souper Subs, 8 out of the 25 employees work part time. What percent of the employees work part time?

    Answer _____

17. Juanita saves 15% of her salary. What fraction of her income does Juanita save?

    Answer _____

18. According to a recent survey, five out of eight young people do some kind of exercise each week. Write this fraction as a decimal.

    Answer _____

Answers begin on page 46.

# Understanding Percent Sentences and Equations

There are three parts to a percent sentence: *the part, the whole,* and the *percent* (also called the *rate*).

You use a percent sentence to make percent equations in order to solve percent problems. In a percent sentence, the word *of* means "multiply", and the word *is* means "equals".

| Percent sentence |
|---|
| 20% of 48 is 9.6. |
| percent  whole  part |

| Percent equation |
|---|
| $0.2 \times 48 = 9.6$ or $\frac{2}{10} \times 48 = 9.6$ |

**Write the percent, the whole, and the part for each problem. Then write the percent equation.**

1. 10% of 30 is 3.

   percent ____10%____

   whole ____30____

   part ____3____

   ____$0.1 \times 30 = 3$____

   25% of 200 is 50.

   percent _____

   whole _____

   part _____

   _____

   2% of 20 is 0.4.

   percent _____

   whole _____

   part _____

   _____

2. 6% of 30 is 1.8.

   percent _____

   whole _____

   part _____

   _____

   400% of 40 is 160.

   percent _____

   whole _____

   part _____

   _____

   $\frac{1}{2}$% of 1,000 is 5.

   percent _____

   whole _____

   part _____

   _____

3. 8% of 22 is 1.76.

   percent _____

   whole _____

   part _____

   _____

   50% of 526 is 263.

   percent _____

   whole _____

   part _____

   _____

   6.5% of 200 is 13.

   percent _____

   whole _____

   part _____

   _____

Answers begin on page 46.

# Finding the Part

Sometimes the part is missing from the percent sentence. To find the part, write a percent equation using $n$ to stand for the part. Change the percent to a decimal or fraction. Multiply to solve the equation.

**What is 92% of 60?**

> Change 92% to 0.92. Write a percent equation using $n$ for the unknown part. Multiply.
> $n = 0.92 \times 60$
> $n = 55.2$

**What is 30% of $40?**

> Change 30% to $\frac{3}{10}$. Write a percent equation using $n$ for the unknown part. Multiply.
> $n = \frac{3}{10} \times \$40$
> $n = \$\frac{120}{10}$
> $n = \$12$

Write a percent equation for each problem. Then solve.

1. What is 50% of 50?   $n = \frac{5}{10} \times 50$     What is 330% of $60?        What is 75% of 80?
                        $n = \frac{250}{10}$
                        $n = 25$

2. 80% of 120 is what number?           What is 1% of 30?           What is 85% of 20?

3. What is 25% of 16?      What is $3\frac{1}{2}$% of $100?           What is 3% of 300?

Change the rate to a decimal. Write a percent equation and solve for the part.

4. 50% of 90   $n = 0.5 \times 90$      300% of $60                  25% of 32
               $n = 45.0$

5. 80% of 120                            75% of $80                  1% of $30

6. 40% of 75                             90% of $200                 $\frac{1}{2}$% of 80

Change the rate to a fraction. Write a percent equation and solve for the part.

7. 25% of 40   $n = \frac{1}{4} \times 40$   200% of $100            30% of $90
               $n = 10$

8. $33\frac{1}{3}$% of $60              $\frac{1}{2}$% of $26        10% of 99

9. 50% of $32                            75% of 50                   $8\frac{1}{2}$% of $120

# Estimating the Part

To estimate the part, round one or both of the numbers in the problem. Change the rounded percent to a decimal or fraction. Write an equation and solve.

**Estimate: 29% of $112.27**

> Round 29% to 30% and $112.27 to $110.00. Change 30% to 0.3. Write an equation and solve.
>
> $n = 0.3 \times \$110$
> $n = \$33$
>
> The exact answer is $32.56, rounded to the nearest cent.

**Estimate: $12\frac{1}{2}$% of 79**

> Round 79 to 80. Change $12\frac{1}{2}$% to $\frac{1}{8}$. Write an equation and solve.
>
> $n = \frac{1}{8} \times 80$
> $n = 10$
>
> The exact answer is 9.88, rounded to the nearest hundredth.

**Estimate. Then find the exact answer. Round answers to the nearest hundredth or cent.**

1. 10% of 20    $n = 0.1 \times 20$      32% of 9                 9% of $53.29
                    $n = 2$

2. 99% of $320                 11% of 182               $4\frac{1}{2}$% of 50

3. $33\frac{1}{3}$% of $11.98          146% of 98.8            48% of 2

**Solve.**

4. Fern wants to find the amount of sales tax she will pay on a purchase of $97.36. Which expression would give the best estimate if the sales tax rate is $4\frac{1}{2}$%?

    a) $0.05 \times \$90$

    b) $0.5 \times \$100$

    c) $0.05 \times \$100$

5. Betty bought a lamp for $29.95. She will sell the lamp in her store with a 29% mark-up. Which expression would give the best estimate of the amount of the mark-up?

    a) $3 \times \$30$

    b) $0.3 \times \$30$

    c) $0.3 \times \$20$

Answers begin on page 46.

# Finding the Part in Percent Word Problems

Estimate. Then find the exact answer. Round answers to the nearest hundredth.

1. A survey at Miller's Textile Mill showed that there are 250 employees. 30% of the employees are men. How many men work at the mill?

   Estimate _____ Answer _____

2. A videocassette recorder that normally sells for $339 is on sale for 25% off. How much will you save during this sale?

   Estimate _____ Answer _____

3. Ed's Lawncare requires a 20% down payment for any jobs over $100. If a job costs $172, what is the amount of the down payment?

   Estimate _____ Answer _____

4. Rachel decided to save 15% of the money she earned. In one month, Rachel earned $703.42. How much money did she save?

   Estimate _____ Answer _____

5. Several years ago Gaylon charged $124.25 to refinish a bathtub. He has raised his price by 28%. How much has the price increased?

   Estimate _____ Answer _____

6. The price of a television is $295.00. The sales tax rate is $7\frac{1}{2}$%. What is the sales tax?

   Estimate _____ Answer _____

7. On an order of 60 glass bottles, 5% were broken when the clerk unpacked them. How many bottles were broken?

   Estimate _____ Answer _____

8. On their vacation Ellen and Tim spent a total of $525. 38% of the money was for the hotel. How much did they spend for the hotel?

   Estimate _____ Answer _____

9. The markup on a pair of athletic shoes is 65%. Run-Fast athletic shoes cost the store $18. How much is the markup?

   Estimate _____ Answer _____

10. Walter read an advertisement for a 30%-off sale on lawn mowers. How much money would Walter save if he bought a lawn mower that normally sold for $219.50?

    Estimate _____ Answer _____

Answers begin on page 46.

# Finding the Percent

To find the percent, first write a percent equation. Use $p$ to stand for percent. To solve, divide the part by the whole. Then change the answer to a percent.

**What percent of $75 is $60?**

> Write a percent equation using $p$. Divide. Change 0.8 to a percent.
>
> $p \times \$75 = \$60$
> $p = \$60 \div \$75$
> $p = 0.8$ or $80\%$

**4 is what percent of 160?**

> Write a percent equation using $p$. Divide. Change 0.025 to a percent.
>
> $p \times 160 = 4$
> $p = 4 \div 160$
> $p = 0.025$ or $2.5\%$ or $2\frac{1}{2}\%$

**Write a percent equation. Find the percent.**

1. What percent of 40 is 8?
    $p \times 40 = 8$
    $p = 8 \div 40$
    $p = 0.20 = 20\%$

    What percent of $25 is $15?

2. What percent of 48 is 7.2?

    What percent of 23 is 6.9?

3. What percent of 180 is 180?

    What percent of $35 is $5.25?

4. What percent of 400 is 268?

    What percent of $140 is $112?

5. What percent of 20 is 15?

    What percent of $40 is $25?

6. What percent of 50 is 1?

    What percent of $40 is $6.50?

7. What percent of 5 is 12?

    What percent of 9 is 54?

8. What percent of 200 is 33?

    What percent of $200 is $12.50?

Answers begin on page 46.

# Estimating the Percent

To estimate the percent, round one or both numbers in the problem to make division easier. Write an equation and solve.

**Estimate: What percent of $91 is $33?**

Round $91 to $90 and $33 to $30.
Write an equation and solve.

$p \times \$90 = \$30$

$p = \dfrac{\$30}{\$90}$

$p = \dfrac{1}{3}$ or $33\dfrac{1}{3}\%$

The exact answer is 36.3% rounded to the nearest tenth.

**Estimate: What percent of 72 is 6.9?**

Round 6.9 to 7 and 72 to 70.
Write an equation and solve.

$p \times 70 = 7$

$p = \dfrac{7}{70}$

$p = 0.1$ or $10\%$

The exact answer is 9.6% rounded to the nearest tenth

---

Estimate. Then find the exact answer. Round answers to the nearest tenth.

1. What percent of $98 is $4.90?

   What percent of 3.5 is 7.2?

2. What percent of 191 is 32?

   What percent of $106 is $1.06?

3. What percent of 215 is 0.27?

   What percent of $14.95 is $0.89?

---

Solve.

4. Jo has a coupon for 50 cents off a large bottle of shampoo that sells for $3.49. Which expression would give the best estimate of the percent savings?

   a) $50 ÷ $3.50
   b) $0.50 ÷ $3.50
   c) $0.50 × $3.50

5. David bought a car for $3,555. He sold the car for $245 less than he paid for it. Which expression would give the best estimate of the percent value the car has lost?

   a) $3,600 ÷ $250
   b) $250 × $3,600
   c) $250 ÷ $3,600

Answers begin on page 46.

# Finding the Percent in Word Problems

Estimate. Then find the exact answer. Round to the nearest tenth of a percent.

1. Arthur bought a used car last year for $2,495. He traded it in on a new car this year and was allowed only $1,250 for it. What percent of its previous value was his old car worth?

   *What percent of $2,495 is $1,250?*

   Estimate _____ Answer _____

2. In the last election, 795 of 820 eligible voters in precinct 1 voted. What percent of eligible voters voted?

   Estimate _____ Answer _____

3. By paying cash, Mr. Terry bought a television for $39.60 off. Otherwise, it would have cost him $395.95. What percent did he save?

   *What percent of $395.95 is $39.60?*

   Estimate _____ Answer _____

4. Thomas paid $7.59 for a football which was originally priced at $12.49. What percent did Thomas pay?

   *What percent of $12.49 is $7.59?*

   Estimate _____ Answer _____

5. On an average day, 12 workers at ViaTech are out sick. If there are a total of 512 workers, what percent are out sick on an average day?

   Estimate _____ Answer _____

6. Dorothy paid $35.95 for a dress which had been marked $39.99. What percent did she pay?

   Estimate _____ Answer _____

7. Virginia made a down payment of $3,850 on a car that cost $10,250. What percent was her down payment?

   Estimate _____ Answer _____

8. Ellen received a $25 per week raise. Her new salary is $347.30 per week. By what percent did her salary increase?

   Estimate _____ Answer _____

9. On his last English test, Gerald got 72 out of 95 questions correct. What percent of the questions did he answer correctly?

   Estimate _____ Answer _____

10. There were 493 students at Kelley School last year. If 72 of the students are new, what percent of students is new?

    Estimate _____ Answer _____

Answers begin on page 46.

# Finding the Whole

To find the whole, change the percent to a decimal or fraction. Write a percent equation using $n$ for the whole. Divide to solve.

**12% of what number is 18?**

Change 12% to 0.12. Write a percent equation using $n$. Solve.
$$0.12 \times n = 18$$
$$n = 18 \div 0.12$$
$$n = 150$$

**$180 is 30% of what number?**

Change 30% to $\frac{3}{10}$. Write a percent equation using $n$. Solve. Invert and multiply.
$$\frac{3}{10} \times n = \$180$$
$$n = \$180 \div \frac{3}{10}$$
$$n = \$180 \times \frac{10}{3} = \frac{\$1{,}800}{3} = \$600$$

**Write a percent equation. Find the whole.**

1. 25% of what number is 17?
$$0.25 \times n = 17$$
$$n = 17 \div 0.25$$
$$n = 68$$

   40 is 32% of what number?

2. 80% of what number is 64?

   135% of what number is $270?

3. 33 is 0.6% of what number?

   45% of what number is 90?

4. 10% of what number is 73?

   $7\frac{1}{2}$% of what number is $1.20?

5. $16 is 25% of what number?

   125% of what number is 5?

6. 200% of what number is $10?

   4.05 is 9% of what number?

7. $62\frac{1}{2}$% of what number is 15?

   100 is 40% of what number?

8. 40 is 40% of what number?

   25% of what number is $10.56?

Answers begin on page 46.

# Estimating the Whole

To estimate the whole, round one or both numbers in the problem. Change the rounded percent to a decimal or fraction. Write an equation and solve.

**Estimate: 48% of what number is $82?**

> Round 48% to 50% and $82 to $80.
> Change 50% to 0.5. Write an equation and solve.
>
> $0.5 \times n = \$80$
> $n = \$80 \div 0.5$
> $n = \$160$
>
> The exact answer is $170.83 rounded to the nearest cent.

**Estimate: 9.96 is 24% of what number?**

> Round 9.96 to 10 and 24% to 25%.
> Change 25% to $\frac{1}{4}$. Write an equation and solve.
>
> $\frac{1}{4} \times n = 10$.
> $n = 10 \div \frac{1}{4}$
> $n = 10 \times \frac{4}{1} = 40$
>
> The exact answer is 41.5.

Estimate. Then find the exact answer. Round answers to the nearest cent or hundredth.

**1.** 67% of what number is 127?

13 is 49% of what number?

**2.** 1 is 21% of what number?

2% of what number is 0.6?

**3.** 112% of what number is 39?

$3.97 is 15% of what number?

Solve.

**4.** Carey saved $9.20 on a pair of jeans at a 25% off sale. Which expression would give the best estimate of the original price of the jeans?

   a) $9 × 0.25
   b) $9 × 0.75
   c) $9 ÷ 0.25

**5.** Al paid $0.56 tax on dinner at a restaurant. The tax rate was 8%. Which expression would give the best estimate of the original price of his dinner?

   a) $0.60 ÷ 0.08
   b) $0.60 × 0.8
   c) $0.60 ÷ 0.8

Answers begin on page 46.

# Finding the Whole in Percent Word Problems

Estimate. Then find the exact answer. Round answers to the nearest hundredth or nearest cent.

1. Filipa bought a pair of shoes on sale for 50% off. She saved $12.70 on the shoes. What was the original price?

   *50% of what number is $12.70?*

   Estimate _____  Answer _____

2. Our school received a discount on books. We bought all of our books for $995. This was 89% of the original price. How much would the books cost without the discount?

   *89% of what number is $995?*

   Estimate _____  Answer _____

3. Mr. Miller sold 82% of his farm. The number of acres sold was 190. How many acres were there in the farm before selling?

   *82% of what number is 190?*

   Estimate _____  Answer _____

4. Pat's Printing has 32 local clients. If 80% of their clients are local businesses, what is the total number of clients?

   Estimate _____  Answer _____

5. Sue Bain paid $5,775 for a car. She paid 68% of the original price because the car had been used as a demonstrator. What was the original price of the car?

   Estimate _____  Answer _____

6. The distance from New York to San Francisco is 5,200 miles by way of the Panama Canal. This is 40% of the distance by way of the Strait of Magellan. How far is it by way of the Strait of Magellan?

   Estimate _____  Answer _____

7. Elise has $362.49 in the savings bank now. This is 58% of the amount she had a year ago. How much did she have a year ago?

   Estimate _____  Answer _____

8. The rug in the living room contains 12 square yards. It covers 72% of the floor. How many square yards of floor surface are there?

   Estimate _____  Answer _____

9. The population of Oak Hill grew to 1,428. This is 119% of the population last year. What was Oak Hill's population last year?

   Estimate _____  Answer _____

10. Joe has gained weight. Joe now weighs 181 pounds. This is 102% of his weight of a year ago. How much did he weigh then?

    Estimate _____  Answer _____

Answers begin on page 46.

# Two-Step Problems: Finding the Original Amount

To find the original amount:

- Find what percent the final amount is of the original amount.
- Divide the final amount by the result to find the original amount.

For sales tax and mark-up problems, add the percent to 100% and divide the final amount by the result.

For discounts or depreciation, subtract the percent from 100% and divide the final amount by the result.

**Find the original amount:**
Final amount: $5.25
Sales tax rate: 5%

> Add the tax rate to 100%.
>
> 100% + 5% = 105%
>
> Divide the final amount by 105%.
>
> $5.25 ÷ 1.05 = $5.00

**Find the original amount:**
Final amount: $225.00
Discount rate: 10%

> Subtract the discount rate from 100%.
>
> 100% − 10% = 90%
>
> Divide the final amount by 90%.
>
> $225 ÷ 0.9 = $250.00

Find the original amount. Round answers to the nearest hundredth or nearest cent.

1. Final amount: $60
   Rate of decrease: $4\frac{1}{2}\%$

   Final amount: $98
   Rate of increase: $12\frac{1}{2}\%$

**Solve.**

2. The price of a sweater after a 10% discount is $9.73. What was the original price?

   Answer _____

3. William paid $22.69 for take-out pizza. The sale tax rate was 8%. What was the price of the pizza alone?

   Answer _____

4. A recent survey in Canton showed that this year an average of 21,000 people took public transportation to work each day. This is a 7% increase from last year. About how many people took public transportation last year? Round your answer to the nearest whole number.

   Answer _____

5. Smithville started a curbside recycling program six months ago. There are now 7,129 families recycling. This is about 40% more than six months ago. About how many families were recycling before the curbside program began? Round your answer to the nearest whole number.

   Answer _____

Answers begin on page 46.

# Two-Step Problems: Finding the Final Amount

To find the final amount:

- Find the amount of increase or decrease by using a percent.
- Add or subtract the result to find the final amount.

For sales tax and mark-up problems, add the amount of increase to the original amount.

For discounts or depreciation, subtract the amount of decrease from the original amount.

**Find the final amount:**
Original amount: $39.00
Sales tax rate: 6%

> Find the amount of increase (amount of tax).
>
> $39.00 × .06 = $2.34
>
> Add the amount of increase (tax) to the original amount.
>
> $39.00 + $2.34 = $41.34

**Find the final amount:**
Original amount: $129.00
Discount rate: 15%

> Find the amount of decrease (amount of discount).
>
> $129.00 × 0.15 = $19.35
>
> Subtract the amount of decrease (discount) from the original price.
>
> $129.00 − $19.35 = $109.65

**Find the final amount. Round answers to the nearest cent or hundredth.**

1. Original amount: $12.50
   Sales tax rate: $7\frac{1}{2}$ %

   Original amount: $11,500
   Discount rate: 7%

**Solve.**

2. Alvin weighed 120 pounds 6 months ago. Since then his weight has increased by 2%. How much does Alvin weigh now?

   Answer _____

3. The population of this town was 1,750. The town grew by 8% over the previous year. What is the current population?

   Answer _____

4. The average number of points scored by the Ridge High basketball team decreased by 8%. They had been averaging 75 points per game. What was their new average?

   Answer _____

5. Last month our utility bill was $98. This month shows a decrease of $12\frac{1}{2}$ %. How much is this month's bill?

   Answer _____

Answers begin on page 46.

# Two-Step Problems: Finding the Percent (Rate) of Increase or Decrease

To find the percent or rate of increase or decrease:

- Subtract to find the amount of increase or decrease.
- Divide to find the percent or rate.

For sales tax and mark-up problems, subtract the original amount from the final amount and divide the result by the original amount.

For discounts or depreciation, subtract the final amount from the original amount and divide the result by the original amount.

**Find the rate of increase:**
Original amount: $24.00
Final amount: $30.00

> Subtract the original amount from the final amount.
>
> $30.00 − $24.00 = $6.00
>
> Divide the result by the original amount.
>
> $6.00 ÷ $24.00 = 0.25 or 25%

**Find the rate of decrease:**
Original amount: $50.00
Final amount: $45.00

> Subtract the final amount from the original amount.
>
> $50.00 − $45.00 = $5.00
>
> Divide the result by the original amount.
>
> $5.00 ÷ $50.00 = 0.1 or 10%

**Find the percent of increase or decrease. Round answers to the nearest tenth of a percent.**

1. Original amount: $9050
   Final amount: $9412

   Original amount: $75
   Final amount: $81

**Solve. Round answers to the nearest hundredth.**

2. Louise bought her condominium for $56,900. After two years, its value has increased to $66,500. What is the percent of increase in the value?

   Answer _____

3. The value of a new car that cost $14,500 decreased to $11,200. What was the percent of decrease in the value of the car?

   Answer _____

4. Last year there were 1,200 employees at City Hospital. This year the staff was cut to 980. What was the percent decrease in staffing at the hospital?

   Answer _____

5. Last semester, the beginning math class at the Adult Center had 15 students. This semester there are 27 students. What was the percent increase in the number of students?

   Answer _____

Answers begin on page 46.

# Simple Interest

To find the *simple* interest, multiply the amount borrowed or invested times the interest rate times the amount of time. Study the example below.

Interest = principal × rate × time

To find the total amount, add the interest to the principal.

Total = principal + interest

**Find the simple interest on a $2,000 loan at 6% for 1 year.**

> Interest = principal × rate × time
> I = prt
> I = $2,000 × 0.06 × 1
> I = $120.00
>
> The total interest is $120.

**Find the total amount required to repay the loan.**

> Total = principal + interest
> Total = $2,000 + $120 = $2,120

**Find the simple interest and the total amount. Round answers to the nearest cent.**

| | | |
|---|---|---|
| 1. $500 is borrowed for 1 year at 6% <br> I = $500 × 0.061 × 1 <br> I = $30 <br> Total = $500 + $30 <br> Total = $530 | $900 is deposited for 1 year at $6\frac{1}{2}\%$ | $1,500 is borrowed for 1 year at 8% |
| 2. $1,000 is deposited for 2 years at $7\frac{1}{2}\%$ | $2,000 is deposited 2 years at 10% | $2,500 is borrowed 2 years at 18% |
| 3. $1,000 is borrowed for $\frac{1}{2}$ year at 5%. | $4,000 is deposited for $\frac{1}{2}$ year at $3\frac{1}{3}\%$ | $300 is borrowed six months at 10% |

**Solve.**

4. Agnes borrowed $5,000 from her mother to buy a car. She will pay 5% simple interest and will repay the loan in 3 years. What is the total amount she will pay?

   Answer _____

5. Dan's bank pays $5\frac{1}{4}\%$ simple interest. How much money will Dan have in his account at the end of a year if he started with a deposit of $350?

   Answer _____

# Compound Interest

*Compound* interest is paid on savings accounts. To find compound interest, first compute the interest, add the interest to the original amount. Then, for the next time period, compute the interest on the total amount. Compound interest may be paid annually, semiannually, monthly or daily.

**Find: Semiannual compound interest on $2,000 at 6% at the end of 1 year.**

Step 1.  I = $2,000 × 0.06 × $\frac{1}{2}$
         I = $60.00
Step 2.  $2,000 + $60.00 = $2060

Step 3.  I = $2,060 × 0.06 × $\frac{1}{2}$
         I = $61.80
Step 4.  $60.00 + $61.80 = $121.80

The total interest is $121.80.
The total amount in the account is $2,000 + $121.80 = $2,121.80.

**Find the amount of compound interest and the total amount in the account. Round to the nearest cent if necessary.**

1. $1,000 is invested for 1 year at 7% compounded semiannually

   Interest _____
   Total _____

   $5,000 is invested for 1 year at $8\frac{1}{2}$% compounded semiannually

   Interest _____
   Total _____

   $1,000 is invested for 1 year at 5% compounded quarterly

   Interest _____
   Total _____

2. $500 is deposited for $1\frac{1}{2}$ years in an account that pays 3% interest compounded semiannually

   Interest _____
   Total _____

   $2,000 is deposited for 1 year in an account that pays $6\frac{1}{4}$% interest compounded semiannually

   Interest _____
   Total _____

   $800 is deposited for 2 years in an account that pays 5% interest compounded annually

   Interest _____
   Total _____

**Solve. Round answers to the nearest cent.**

3. Agnes has had $350 in a savings account for a year. She gets $5\frac{1}{2}$% interest compounded semiannually. How much total interest will she receive at the end of a year?

   Answer _____

4. Jessica's bank pays $4\frac{1}{2}$% interest compounded semiannually. If she deposits $400 in a savings account, how much will she have in the account in 1 year?

   Answer _____

Answers begin on page 46.

# Checking Up on Percents

**Change to a percent.**

1. 0.25 = _____    0.01 = _____    1.2 = _____

2. 0.9 = _____    2.6 = _____    0.435 = _____

3. $\frac{1}{5}$ = _____    $\frac{3}{4}$ = _____    $\frac{2}{3}$ = _____

**Change to a decimal and to a fraction.**

4. 7% = _____    99% = _____    102% = _____

5. $\frac{1}{2}$% = _____    4% = _____    0.4% = _____

6. 1% = _____    150% = _____    62% = _____

**Find each answer. Round answers to the nearest cent or nearest hundredth.**

7. What is 90% of 60?    What percent of $50 is $25?

8. 10% of what number is $7?    What percent of 2 is 4?

9. What is 150% of 6?    25% of what number is 80?

10. What is 0.3% of 30?    8% of what number is $10?

11. What percent of $19.60 is $0.98?    What is $6\frac{1}{2}$% of $28.40?

**Estimate. Then find the exact answer. Round answers to the nearest cent or nearest hundredth.**

12. What is 39% of 219?    What percent of $82 is $9?

13. $200 is 25% of what number?    What is $66\frac{2}{3}$% of 17?

14. 63 is what percent of 127?    16% of what number is 22?

42    Answers begin on page 46.

# Checking Up on Percents

Estimate. Then find the exact answer. Round answers to the nearest cent or nearest hundredth.

---

**15.** The price of a new stereo is $199.50. The sales tax rate is 6%. What is the amount of tax?

Estimate _____ Answer _____

**16.** Barry paid $13.27 for a shirt that was originally priced at $20.00. What percent did Barry save?

Estimate _____ Answer _____

---

**17.** Mary gained weight over the holidays. She now weighs 132 pounds. This is 10% more than her previous weight. How much did she weigh before the holidays?

Estimate _____ Answer _____

**18.** Enrollment is up 25% in our town's schools. Last year's enrollment was 1,092. What is the present enrollment?

Estimate _____ Answer _____

---

**19.** The average number of points scored by one college football team decreased by 11%. The team had been averaging 23 points per game. What is their new average?

Estimate _____ Answer _____

**20.** Warren bought a house for $52,500. In two years its value increased to $60,000. What is the percent of increase?

Estimate _____ Answer _____

---

**21.** The population of one town is now 23,400. This is an increase of 100% since last year. What was the population last year?

Estimate _____ Answer _____

**22.** Zachary borrowed $4500 from his brother to buy a boat. He will pay 7% simple interest and will repay the loan in 2 years. What is the total amount he will pay?

Estimate _____ Answer _____

---

**23.** Jorg's bank pays 5% interest compounded semiannually. If he deposits $500 in a savings account, how much will he have in the account in 1 year?

Estimate _____ Answer _____

**24.** The distance from New York to San Francisco is 5,200 miles by way of the Panama Canal. This is 40% of the distance by way of the Strait of Magellan. How far is it by way of the Strait of Magellan?

Estimate _____ Answer _____

---

Answers begin on page 46.

# Progress Review

**Write the following in words.**

1. 0.26 _____     2.09 _____

2. $32.60 _____     $0.15 _____

**Write each decimal as a fraction or a mixed number. Write each fraction as a decimal.**

3. 0.8 _____   $\frac{1}{10}$ _____   1.02 _____   $\frac{32}{10}$ _____

**Compare. Write <, >, or =.**

4. 0.2 ____ $\frac{2}{10}$    3.006 ____ $\frac{36}{100}$    15.27 ____ 15.207    $\frac{289}{1000}$ ____ 2.89

**Round as indicated.**

5. 8.63 (nearest whole number) _____    $31.50 (nearest dollar) _____

6. $1.099 (nearest cent) _____    0.496 (nearest tenth) _____

7. 5.0375 (nearest hundredth) _____    $0.86 (nearest dime) _____

**Estimate by rounding as indicated. Then find the exact answer.**

8. (nearest tenth)        (nearest dollar)       (nearest dime)

   2.79 + 5.61 =          $5.98 × 24 =           2)$27.78 =

9. (nearest hundredth)    (nearest cent)         (nearest whole number)

   18.062 − 5.097 =       $9.382 × 10 =          8.3)447.9

**Add, subtract, multiply, or divide.**

10.  $2.66        $8.59         $10.00        0.23
    + 3.48       × 0.08        −  1.93        0.597        1.5)45
                                              + 1.6

**Solve.**

11. Juanita saves 15% of her salary. What fraction of her salary does Juanita save?

12. A copper wire 48 feet long was divided into 15 pieces. How long was each piece?

Estimate _____ Answer _____    Estimate _____ Answer _____

**44**    **Answers begin on page 46.**

# Progress Review

Find each answer. Round answers to the nearest cent or nearest hundredth.

13. What is 2% of 200?

    What percent of $90 is $30?

14. 85% of what number is $51?

    What is 110% of 100?

15. What percent of 6 is 24?

    0.2% of what number is 1?

Estimate. Then find the exact answer. Round answers to the nearest cent or nearest hundredth.

16. What is 61% of 57?

    What percent of $96.75 is $4.95?

17. $11 is 32% of what number?

    What is $33\frac{1}{3}$% of 29?

18. 82 is what percent of 101?

    71% of what number is $1.29?

---

19. Jackie paid $316.40 for a couch. The sales tax rate was 8%. What was the original price of the couch?

    Estimate _____ Answer _____

20. Last year Sheila's salary was $17,500. This year her salary is $18,900. What percent increase is this?

    Estimate _____ Answer _____

21. John planted 15 acres of corn when he first started farming. He has increased the number of acres he plants by 27%. How many acres does he plant now?

    Estimate _____ Answer _____

22. Sally spends 12% of her monthly salary for food. If she makes $1,250 each month, how much does she spend each month on food?

    Estimate _____ Answer _____

23. Mr. Stone paid $63.56 for a plane ticket. Two days later the airline offered a discount of 50%. How much would he pay for the ticket now?

    Estimate _____ Answer _____

24. Nancy borrowed $200 for 6 months to buy gifts. The loan company charged 18% simple interest. How much will she repay?

    Estimate _____ Answer _____

Answers begin on page 46.

# ANSWER KEY

## Page 2
1. 0.3; three tenths
   0.1; one tenth
   2.4; two and four tenths
2. 0.32; thirty-two hundredths
   0.48; forty-eight hundredths
   1.15; one and fifteen hundredths
3. 0.95; ninety-five hundredths
   0.01; one hundredth
   1.05; one and five hundredths

## Page 3
1. 0.3    0.15
2. 0.025    0.005
3. 0.0015    0.3 feet
4. 0.5 yard    0.75 mile
5. 0.015 inch    0.04 mile
6. 0.20 ton    0.205
7. 18.9    10.65
8. 2.1    71.06
9. one tenth
   twenty-five hundredths
10. six hundredths
    sixty-two hundredths
11. three hundred twelve thousandths
    seven thousandths
12. five tenths
    sixteen hundredths
13. thirty-seven thousandths
    nine hundred two thousandths
14. three and forty-nine hundredths
    thirty-five and two tenths
15. seventeen and nine tenths
    fifty and five hundredths
16. one and four tenths
    ninety-six and six hundred twenty-one thousandths

## Page 4
1. $0.52    $0.18
2. $0.99    $0.70
3. $1.05    $5.10
4. $12.19    $23.50
5. $62.30    $82.09
6. $136.92
7. eleven cents
   eighty cents
8. forty-four cents
   five cents
9. nine cents
   seventy-one cents
10. five dollars and thirteen cents
    seven dollars and one cent
11. ten dollars and ten cents
    fifteen dollars and twenty-two cents
12. thirty-nine dollars and ninety eight cents
    fifty-two dollars and thirty-eight cents
13. seventy dollars and three cents
    eighty-nine dollars and forty-seven cents
14. one hundred twenty-three dollars and fifty-two cents
15. five hundred eighty dollars and fifty cents
16. four hundred twenty-seven dollars and nine cents

## Page 5
1. $\frac{2}{5}$    $\frac{3}{5}$    $\frac{2}{25}$    $\frac{1}{500}$
2. $\frac{21}{100}$    $\frac{83}{1,000}$    $\frac{901}{1,000}$    $\frac{9}{500}$
3. $4\frac{1}{2}$    $1\frac{31}{50}$    $10\frac{1}{10}$    $1\frac{11}{40}$
4. $9\frac{7}{100}$    $38\frac{6}{25}$    $5\frac{23}{50}$    $13\frac{4}{5}$
5. 0.1    0.2    0.5    0.7
6. 0.06    0.80    0.052    0.416
7. 5.6    3.1    7.6    0.65
8. 1.03    5.09    1.643    2.051

## Page 6
1. =    =    <    <
2. =    =    <    >
3. =    =    <    <
4. $0.25, $1.01, $1.02, $1.10, $1.20, $1.25, $2.50, $25
5. D    D
6. S    D
7. D    S
8. D    S
9. D    S    D    D
10. D    S    S    D
11. D    D    S    S
12. D    D    S    D
13. 9,999.9
14. 10,000.0

## Page 7
1. 6    2    1    315    19
2. $28.00    $520.00    $183.00
   $92.00    $6.00
3. 244    1    0    119    129
4. $0.70    $0.50    $2.80
   $15.00    $32.10
5. 29.3    33.3    92.1    5.4    15.6
6. 12.2    0.6    $166.70
   $1.10    423.8
7. 1.53    3.04    0.38
   8.67    0.33
8. $10.08    $18.28    $5.90
   $27.99    $1.33
9. 435.33    $0.55    $11.10
   183.27    136.37

## Page 8
1. 20    $13.00    465    4
2. $4.70    101.6    $29.50    16.525
3. 0.1    102.2    $11.00    1.1
4. c
5. b

## Page 9
1. $9.57    $9.35    $0.34
   22.53    20.40
2. 34.012    6.166    77.663
   87.78    8.239
3. 3.75    10.395    2.709
   $8.70    7.199
4. $14.30    $10.22    $1.42
5. $42.57    $149.07
6. 167.30    $210.83    $8.38
   $0.59

## Page 10
1. $5.12    $4.74    $2.15
   2.014    0.160
2. $6.96    $0.05    $24.50
   $79.78    $9,032.30
3. 1.625    $3.87    4.326
   $1.09    $15.30
4. 3.834    $1.51    $73.49
5. $5.50    1.04    $0.53
6. $83.70    4.2    $13.00
   3.1

## Page 11
1. $6.19
2. 5.77 minutes
3. $5,225.00
4. 1,550.4 miles
5. 19.5 hours
6. 0.047 feet
7. 0.63 second
8. 6.625 inches
9. $8.54
10. 0.343

## Page 12
1. three and twenty-nine hundredths
   seventeen and six thousandths
2. fourteen dollars and fifty cents
   sixty-seven cents
3. 9/10    4 1/100    0.8    1.3
4. >    =    <    >
5. 24    $8.00
6. $0.38    18.5
7. 0.52    $93.90
8. 4    $21.00    $13.30
9. 1.88    $11.08    144
10. $2.79    7.21    $52.09
    8.466    $690.85
11. 13.963    $25.44    $267.25
    1.263    $3,858.21
12. $8.07    $45.13    $6.80
13. $31.26    $3.33    $21.00

## Page 13
1. 1.724    0.252    $9.78
   14.4    $11.70
2. $27.20    $114.08    $33.90
   $100.02    $130.00
3. 98.4    126.7    $350.00
   $0.20    49.5
4. $76.75    11.7    3.69
5. 144    332    $300    $350

## Page 14
1. 0.18    0.018    $1.12
   4.92    0.075    $0.25
2. $0.03    0.0081    0.0024
   0.00048    $1.10    $0.01
3. 0.0021    0.0165    $0.06
   $1.86    3.706    $3.81
4. $38.25    $1.39    $11.99
5. $1.47    $8.56    416
   0.45

## Page 15
1. 8.2    $0.69    $2.76    0.112
2. 0.04    $1.50    0.203    $3.14
3. 41.6    $41.20    1.8    0.7
4. $53.00    12    $12.00
5. $8.70    1    $181.80    15,070

## Page 16
1. 0.5    $0.30    $0.40
   0.02    $0.75
2. 0.02    0.05    $0.67
   $0.25    $0.95
3. 0.25    0.125    0.3
   0.6    0.1875
4. 0.5    0.625    0.2
   0.4    0.75

## Page 17
1. 75    $460.00    0.7
2. 70    $460.00    7.5
3. 5    800    $125.00
4. 1,250    125    $1,492.00
5. 6,200    $6,421.50    $64,215.00
6. 0.00315    0.00048    0.000048
7. 0.0375    $0.375    3.75
8. $37.50    0.000375    0.000007
9. $7.1935    0.016147    $0.01492
10. $2.6718    0.0026718    0.026718

## Page 18
1. $1,530.00
2. 145.5 miles per hour
3. 908 miles
4. 8.60565 pounds
5. 64 miles
6. 102.9952 kilometers
7. $8.50
8. 3.2 acres
9. 22.7 miles per gallon
10. 26.5 bushels per acre

## Page 19
1. $5.94    1.131    $4.12    0.684732    $232.56
2. $4.35    2.52    $268.56    $31.35    1.102
3. $0.58    0.11    3.12    $2.33    45
4. $2.71    $0.50    950    69.07
5. $0.07    $47.98    $3.97    1.21
6. $3.95    107    6
7. 11.26    0.049    1,230

## Page 20
1. one hundredth
   three and twenty-four hundredths
2. twenty-five dollars and thirty cents
   nineteen cents
3. $\frac{23}{100}$    0.4    $6\frac{9}{100}$    3.25
4. >    >    <    =
5. 32    $43.00
6. $0.43    7.4
7. 0.48    $8.50
8. 2.1    $56.00    $35.00
9. 4.16    $0.96    43
10. $3.38    $7.29    $0.05    $18.82    12
11. 17.627    36.96    0.3168    $0.50    2.55
12. 0.12    $259.64    $0.46    0.0045    0.06

## Page 21
13. $6.23    84    1,200    6.309    0.9996
14. 0.166    0.227    0.014    $4.54    150
15. 36    $2,930.00    0.16
16. $90.00    $0.06    0.246
17. 0.01    $46.10    $32.00
18. 19.5 miles per gallon
19. $175.63
20. $423.85
21. $3.56
22. 3.2 feet
23. 6,687.2 pounds

## Page 22
1. $\frac{5}{100}$ or 5%    $\frac{25}{100}$ or 25%
   $\frac{36}{100}$ or 36%    $\frac{10}{100}$ or 10%
2. 40% shaded, 60% unshaded, 100% total
   90% shaded, 10% unshaded, 100% total
   50% shaded, 50% unshaded, 100% total
   75% shaded, 25% unshaded, 100% total
3. $\frac{5}{100}$     $\frac{95}{100}$

   $\frac{27}{100}$     $\frac{75}{100}$
4. $\frac{5}{100}$, 5%    $\frac{25}{100}$, 25%
   $\frac{50}{100}$, 50%    $\frac{10}{100}$, 10%
   $\frac{100}{100}$, 100%

## Page 23
1. 0.1%, 1%, 10%, 100%
2. 0.5%, 5%, 50%, 500%
3. 0.25%, 2.5%, 25%, 250%
4. 0.75%, 7.5%, 75%, 750%
5. 5%    8. 150%
6. 100%    9. 50%
7. 10%    10. 2.5%

## Page 24
1. 30%    0.8%    245%    375%
2. 91%    150%    1.5%    0.04%
3. 73%    67.2%    2%    67%
4. 325%    9%    70%    2,393%
5. 37.5%    200%    70%
6. 21%    312.5%    25%
7. 60%    85%    0.2%
8. $66\frac{2}{3}$%    $11\frac{1}{10}$%    $41\frac{2}{3}$%

## Page 25
1. 0.82, $\frac{41}{50}$    0.07, $\frac{7}{100}$
2. 0.01, $\frac{1}{100}$    1.42, $1\frac{21}{50}$
3. 0.095, $\frac{19}{200}$    0.55, $\frac{11}{20}$
4. 1.09, $1\frac{9}{100}$    0.73, $\frac{73}{100}$
5. 0.98, $\frac{49}{50}$    0.004, $\frac{1}{250}$
6. 0.04, $\frac{1}{25}$    0.44, $\frac{11}{25}$
7. 1.75, $1\frac{3}{4}$    0.83, $\frac{83}{100}$
8. 0.26, $\frac{13}{50}$    0.005, $\frac{1}{200}$
9. 0.003, $\frac{3}{1,000}$    1.27, $1\frac{27}{100}$
10. 0.125, $\frac{1}{8}$    0.95, $\frac{19}{20}$

11. 0.0005, $\frac{1}{2,000}$    20, $\frac{20}{1}$
12. 0.065, $\frac{13}{200}$    5, $\frac{5}{1}$

## Page 26
1. $\frac{1}{10}$ = 0.1 = 10%
2. $\frac{13}{20}$ = 0.65 = 65%
3. $\frac{23}{100}$ = 0.23 = 23%
4. $\frac{17}{20}$ = 0.85 = 85%
5. $\frac{1}{2}$ = 0.5 = 50%
6. $\frac{3}{4}$ = 0.75 = 75%
7. $\frac{300}{100}$ = 3 = 300%
8. $1\frac{9}{10}$ = 1.9 = 190%
9. $2\frac{1}{2}$ = 2.5 = 250%
10. $\frac{1}{3}$ = $0.33\frac{1}{3}$ = $33\frac{1}{3}$%
11. $\frac{2}{3}$ = $0.66\frac{2}{3}$ = $66\frac{2}{3}$%
12. $\frac{1}{200}$ = 0.005 = $\frac{1}{2}$%
13. 21.875%    16. 32%
14. 0.075    17. $\frac{3}{20}$
15. 0.1    18. 0.625

## Page 27
1. 10%
   30
   3
   0.1 x 30 = 3
   25%
   200
   50
   0.25 x 200 = 50
   2%
   20
   0.4
   0.02 x 20 = 0.4
2. 6%
   30
   1.8
   0.06 x 30 = 1.8
   400%
   40
   160
   4 x 40 = 160
   $\frac{1}{2}$%
   1,000
   5
   0.005 x 1,000 = 5
3. 8%
   22
   1.76
   0.08 x 22 = 1.76
   50%
   526
   263
   0.5 x 526 = 263
   6.5%
   200
   13
   0.065 x 200 = 13

**Page 28**
1. 25    $198.00    60
2. 96    0.3        17
3. 4     $3.50      9
4. 45.0  $180.00    8
5. 96    $60        $0.30
6. 30    $180.00    0.4
7. 10    $200.00    $27.00
8. $20.00 $0.13     9.9
9. $16.00 37.5      $10.20

**Page 29**
1. 2       2.88    $4.80
2. $316.80 20.02   2.25
3. $3.99   144.25  0.96
4. c
5. b

**Page 30**
1. 75 men      6. $22.13
2. $84.75      7. 3 bottles
3. $34.40      8. $199.50
4. $105.51     9. $11.70
5. $34.79      10. $65.85

**Page 31**
1. 20%     60%
2. 15%     30%
3. 100%    15%
4. 67%     80%
5. 75%     62.5%
6. 2%      16.25%
7. 240%    600%
8. 16.5%   6.25%

**Page 32**
1. 5%      205.7%
2. 16.8%   1%
3. 0.1%    6%
4. b
5. c

**Page 33**
1. 50.1%    6. 89.9%
2. 97%      7. 37.6%
3. 10%      8. 7.8%
4. 60.8%    9. 75.8%
5. 2.3%     10. 14.6%

**Page 34**
1. 68       125
2. 80       $200.00
3. 5,500    200
4. 730      $16.00
5. $64.00   4
6. $5.00    45
7. 24       250
8. 100      $42.24

**Page 35**
1. 189.55   26.53
2. 4.76     30
3. 34.82    $26.47
4. c
5. a

**Page 36**
1. $25.40
2. $1,117.98
3. 231.71 acres
4. 40 clients
5. $8,492.65
6. 13,000 miles

7. $624.98
8. 16.67 square yards
9. 1,200
10. 177.45 pounds

**Page 37**
1. $62.83    $87.11
2. $10.81
3. $21.01
4. 19,626
5. 5,092

**Page 38**
1. $13.44    $10,695.00
2. 122.4 pounds
3. 1,890 people
4. 69 points per game
5. $85.75

**Page 39**
1. 4% increase    8% increase
2. 16.87 increase
3. 22.76% decrease
4. 18.33% decrease
5. 80% increase

**Page 40**
1. I = $30, T = $530
   I = $58.50, T = $958.50
   I = $120, T = $1,620
2. I = $150, T = $1,150
   I = $400, T = $2,400
   I = $900 T = $3,400
3. I = $25, T = $1,025
   I = $66, T = $4,066
   I = $15, T = $315
4. $5,750
5. $368.38

**Page 41**
1. $71.23, $1,071.23
   $434.03, $5,434.03
   $50.94, $1,050.94
2. $22.84, $522.84
   $126.95, $2,126.95
   $82.00, $882.00
3. $19.52
4. $418.20

**Page 42**
1. 25%     1%      120%
2. 90%     260%    43.5%
3. 20%     75%     66%
4. 0.07, $\frac{7}{100}$    0.99, $\frac{99}{100}$    1.02, $1\frac{1}{50}$
5. 0.005, $1\frac{1}{200}$    0.04, $\frac{1}{25}$    0.004, $\frac{1}{250}$
6. 0.01, $\frac{1}{100}$    1.5, $1\frac{1}{2}$    0.62, $\frac{31}{50}$
7. 54       50%
8. $70      200%
9. 9        320
10. 0.09    $125
11. 5%      $1.85
12. 85.41   11%
13. $800    11.33
14. 49.61%  137.5

**Page 43**
15. $11.97        20. 14.29%
16. 33.65%        21. 11,700
17. 120 pounds    22. $5,130
18. 1,365         23. $525.31
19. 20.47 points  24. 13,000 miles

**Page 44**
1. twenty-six hundredths
   two and nine hundredths
2. thirty-two dollars and sixty cents
   fifteen cents
3. $\frac{4}{5}$    0.1    $1\frac{1}{50}$    3.2
4. =     >     >     <
5. 9         $32
6. $1.10     0.50
7. 5.04      $0.90
8. 8.4       $143.52    $13.89
9. 12.965    $93.82     53.96386
10. $6.14    $0.6872    $8.07
    2.427    30
11. $\frac{3}{20}$
12. 3.2 feet

**Page 45**
13. 4         $33\frac{1}{3}\%$
14. 60        110
15. 400%      500
16. 34.77     5.12%
17. $34.38    9.67
18. 81.19%    $1.82
19. $292.96
20. 8%
21. 19.05 acres
22. $150
23. $31.78
24. $218